SIDNEY KETTLE

Sidney Kettle graduated from the University of Leeds. After obtaining his PhD in the Inorganic Chemistry Department at Cambridge and spending two years in the Theoretical Chemistry Department there. he joined the staff at Sheffield. In 1970 he took up his present appointment of Professor of Inorganic and Theoretical Chemistry at the University of East Anglia.

CHEMISTRY CASSETTES

General Editor:
Peter Groves
The University of Aston in Birmingham

Published 1976 by the Educational Techniques Subject Group, The Chemical Society, London.

© Sidney Kettle 1976

ISBN 0 85186 729 4

The views expressed in this booklet and in the associated tape recording are those of the author and not necessarily those at the Chemical Society.

USING CHEMISTRY CASSETTES

Please read this carefully before you start

This Chemistry Cassette presentation comprises two audio-cassettes with this accompanying workbook. They are designed to be used together and you should have the workbook in front of you as you listen to the cassettes. Material in the workbook is divided into numbered *frames* and Professor Kettle frequently refers to these as he speaks. Each frame contains diagrams, tables and other relevant material and you should locate and study this wherever appropriate.

During the course of the presentation you are asked, from time to time, to stop the tape and to work on some problems: you should, therefore, also have pencil and paper with you. Two of the problems ask you to make observations on a small cube and on a model of the water molecule. You should have these ready before you start. For the cube, a child's building brick would be suitable provided that all the faces are the same colour. Instructions for making a cube from a piece of card are given in frame 10. For the water molecule (which is considered at the start of the second cassette) a simple ball and stick model would be quite adequate. If this is not available, a model made from three balls of plasticine and two matchsticks would be equally suitable.

An important feature of tape recorded material is that it is 'self-pacing'. This means that you can work through it at your own pace, switching off the player whenever you wish to pause for thought, to study a diagram, to work on a problem, etc., and you can use the rewind control on the player to repeat material that you may not have fully understood on a first hearing. To gain the greatest benefit from this presentation you should make full use of these features. You should also make appropriate notes to supplement the material contained in the workbook.

Part	Side	Approximate running times	Corresponding frame numbers
1	A	20 mins.	1 − 10
	B	21 mins.	11 − 28
2	A	38 mins.	29 − 44
	B	28 mins.	45 − 58

PART 1

1

The ground to be covered in this presentation

Symmetry elements

Symmetry operations

Multiplication of symmetry operations

Irreducible Representations of a group

Character tables

Reducible representations of a group

Example: The vibrations of the water molecule

Selection Rules

Molecular Integrals

FRAME CONTINUED ON NEXT PAGE

1
CONTD.

Some (simplified) definitions

A symmetry element   A physical manifestation of the existence of symmetry; for instance, a rotation axis or mirror plane

A symmetry operation: The act of carrying out the operation implied by the existence of a symmetry element. For instance, the act of rotating or the act of reflecting.

In mathematical group theory (not dealt with in this treatment) symmetry operations are represented by matrices

Multiplication   The product of two symmetry operations is the single operation which produces the same end result as the two symmetry operations acting the one after the other. Usually given in the form of a table.

A Group   of symmetry operations is composed of all of the distinct symmetry operations which may occur in a multiplication table.

A Character table   A (square) table which contains numbers, usually integers, which, individually, characterise the behaviour of an object under a symmetry operation. The rows of a character table are called irreducible representations.

Reducible representations   Are sets of numbers which may be written as a sum of irreducible representations. Although not discussed in the present treatment, corresponding to each reducible representation is a set of matrices.

Direct Products   When the corresponding characters of two irreducible representations of a group are multiplied together (arithmetically) a representation of the group is obtained which is the direct product of the two irreducible representations which were multiplied together. Usually given in the form of a table.

2

A <u>symmetry element</u> is a physical manifestation of
the existence of symmetry in an object. Examples
of symmetry elements are rotation axes, mirror planes and
a centre of symmetry

3

The five types of symmetry element are:-

1)   Rotation Axes

2)   Mirror Planes

3)   A centre of symmetry

4)   Rotation reflection axes of symmetry

       (crystallographers prefer to call these

       rotation inversion axes)

5)   The identity element

4

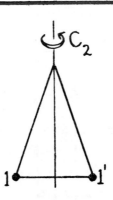

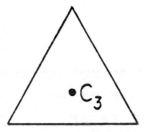

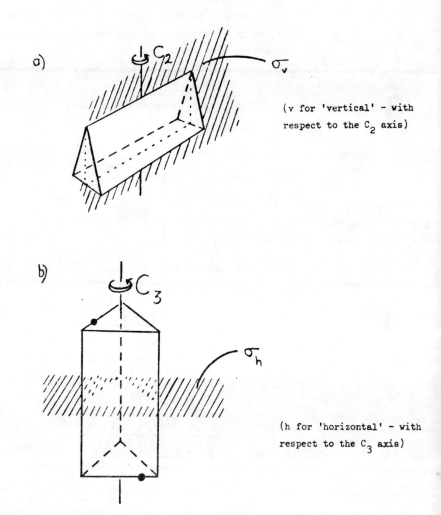

a)

C_2

σ_v

(v for 'vertical' - with respect to the C_2 axis)

b)

C_3

σ_h

(h for 'horizontal' - with respect to the C_3 axis)

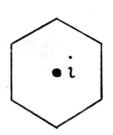

As shown in the bottom
diagram, an arbitrary
line drawn through a centre
of symmetry cuts the figure
at equivalent points
(+, * and ✖) on either side
of the centre of symmetry

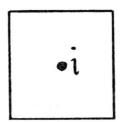

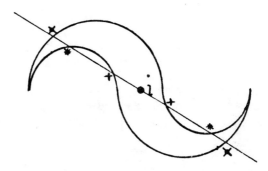

7

An S_4 axis: consider the S_4 axis of allene:-

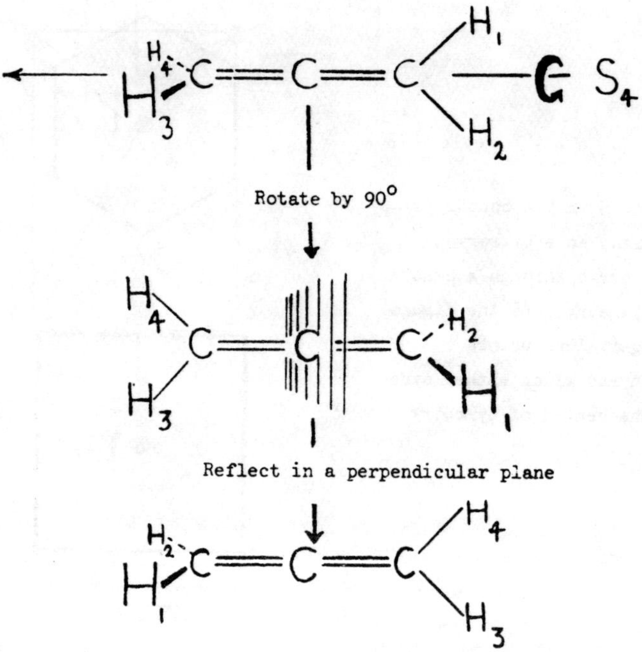

$$H_4\ H_3 C = C = C \overset{H_1}{\underset{H_2}{\diagdown}} \quad \longleftarrow \quad S_4$$

Rotate by $90°$

$$H_4\ H_3 C = C = C \overset{H_2}{\underset{H_1}{\diagdown}}$$

Reflect in a perpendicular plane

$$H_2\ H_1 C = C = C \overset{H_4}{\underset{H_3}{\diagdown}}$$

FRAME CONTINUED ON NEXT PAGE

7
CONTD.

An S_6 axis

rotate $60°$

reflect in a
plane
perpendicular
to the axis of
rotation

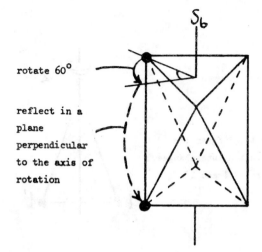

8

A <u>symmetry operation</u> is most simply thought
of as the act of carrying out the operation
implied by the existence of a symmetry element.
However, symmetry operations are more pertinent to the
symmetry aspects of chemistry than are symmetry elements.
This is because an algebra can be constructed associated with
symmetry operations but not with symmetry elements. We
touch on some aspects of this algebra in the
present treatment.

9

In the top diagram the C_2 operation interchanges the corners 1 and 1'

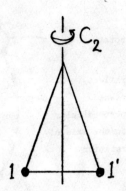

In the bottom diagram the i operation turns 1 into A (and 1' into B). The σ_h turns A into 1' (and B into 1). It follows that C_2 is equivalent to i followed by σ_h.

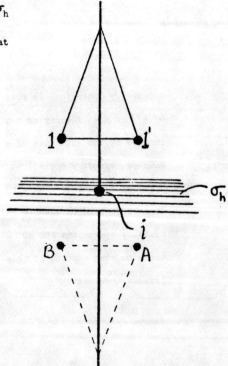

10

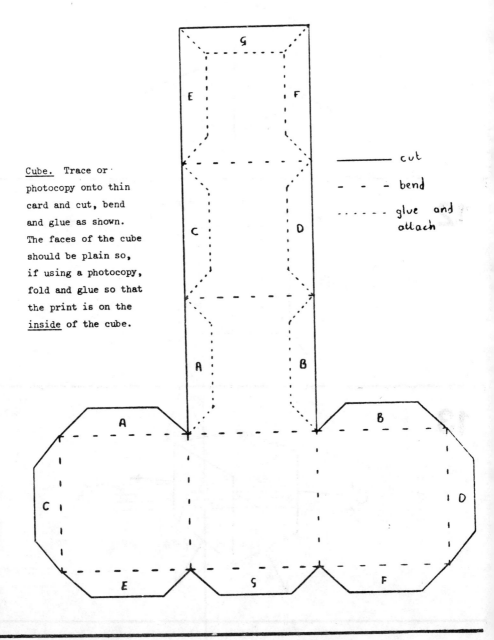

Cube. Trace or
photocopy onto thin
card and cut, bend
and glue as shown.
The faces of the cube
should be plain so,
if using a photocopy,
fold and glue so that
the print is on the
inside of the cube.

—————— cut

– – – bend

· · · · · glue and attach

11

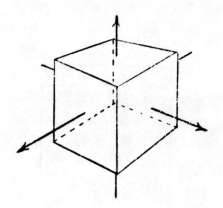

12

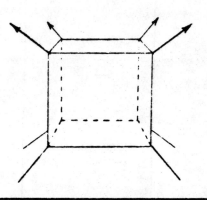

13

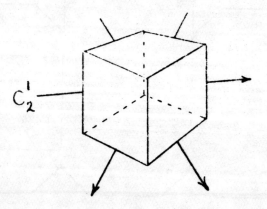

14

A <u>proper rotation</u> is a pure rotation operation. Examples are rotation by $360°$ $(=C_1=E)$, by $180°$ $(=C_2)$, by $120°$ $(=C_3)$, by $90°$ $(=C_4)$, by $72°$ $(=C_5)$, by $60°$ $(=C_6)$ and by $51.43°$ $(=C_7)$.

An <u>improper rotation</u> is a pure rotation combined with (i.e. preceded or followed by) the inversion operation. Examples are C_1 followed by i $(=i)$, C_2 followed by i $(=\sigma)$, C_3 followed by i $(=S_3)$, C_4 followed by i $(=S_4)$ and C_5 followed by i $(=S_5)$. Note that this definition of S_n axes is in accord with the practice of crystallographers (see Frame 3).

15

The i operation

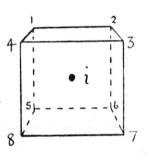

interchanges the corners 1 and 7
2 and 8
3 and 5
4 and 6

to give

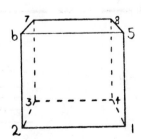

Before the S_4 operation

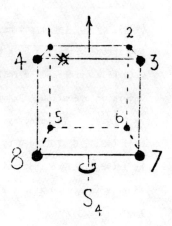

S_4

After

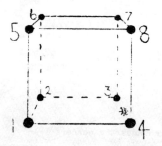

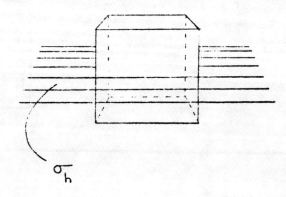

σ_h

20

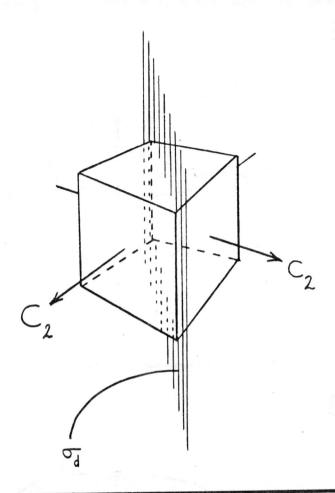

C_2

C_2

σ_d

21

Proper rotations of a cube

$$E \quad 4C_3 \quad 3C_4 \quad 3C_2 \quad 6C_2{}'$$

Improper rotations of a cube

$$i \quad 4S_6 \quad 3S_4 \quad 3\sigma_h \quad 6\sigma_d$$

22

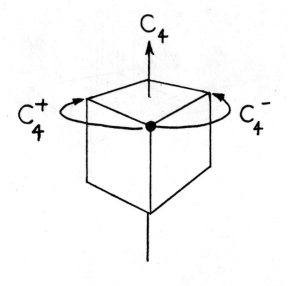

23

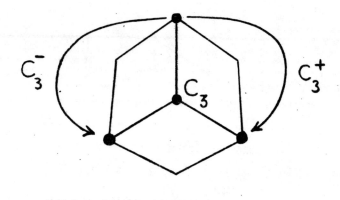

24

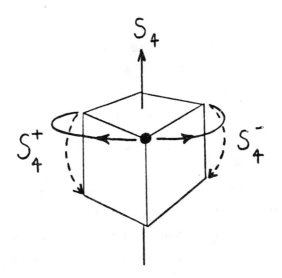

25

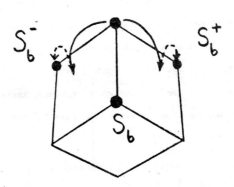

Operations which are in the same _class_ are either derived from a common symmetry element or derived from a set of equivalent symmetry elements. Thus, in the C_{3v} point group the classes are E, $2C_3$ and $3\sigma_v$. The two C_3 operations are derived from a common symmetry element and the three σ_v operations are derived from a set of equivalent symmetry elements.

This is a practical, not mathematical, definition and covers almost all cases commonly encountered. That it is not a perfect definition is seen from Frame 33 where, for instance, in the C_3 point group the operations C_3 and C_3^2 (the latter may be thought of either as a rotation of $240°$ or of $120°$ in the opposite direction to that of the C_3 rotation. In the former case we write the two operations as C_3 and C_3^2; in the latter C_3^+ and C_3^-) fall into different classes. A rather better (but still not perfect) definition is that two operations are in the same class when there exists within the group some _third_ operation which when combined with one gives the other. Thus, in the C_{3v} point group, a C_3 rotation followed by a (correctly chosen) σ_v gives the same nett effect as a single C_3^2 operation. Hence C_3 and C_3^2 ($\equiv C_3^+$ and C_3^-) are in the same class. In the C_3 point group, however, there exists no third operation which may be combined with C_3 to give C_3^2 (or, equivalently, with C_3^+ to give C_3^-).

The correct definition of class involves a fourth operation (F) which has the property that when combined with the third (T) it gives the identity i.e. it 'undoes' the effect of the third operator (T). Two operators A and B are in the same class if a T (and F) can be chosen such that:-

T followed by A followed by F gives the same effect as B on its own.

With this definition T can be _any_ operation in the group (including A or B).

E	$8C_3$	$6C_4$	$3C_2$	$6C_2'$
i	$8S_6$	$6S_4$	$3\sigma_h$	$6\sigma_d$

Note that the total number of operations

$(1 + 8 + 6 + 3 + 6 + 1 + 8 + 6 + 3 + 6) = 48$

is exactly divisible by the number of operations

in any class : $\frac{48}{8} = 6$, $\frac{48}{6} = 8$, $\frac{48}{3} = 16$.

The total number of operations in a group is called

the ORDER of the group. The symmetry operations of

a cube comprise a group of order forty-eight.

28

<u>The ground to be covered in this presentation</u>

Symmetry elements

Symmetry operations

Multiplication of symmetry operations

Irreducible Representations of a group

Character tables

Reducible representations of a group

Example: The vibrations of the water molecule

Selection Rules

Molecular Integrals

PART 2

29

The four elements are

1) The identity (leave alone)

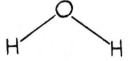

2) A C_2 rotation

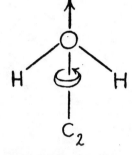

3) Reflection in a σ_v mirror plane (in the plane of the paper)

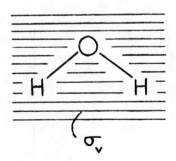

4) Reflection in a second type of σ_v mirror plane (denoted σ_v') perpendicular to the plane of the paper

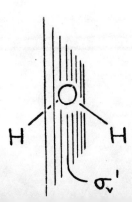

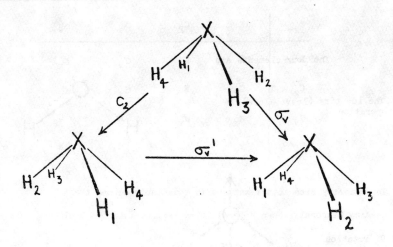

Second operation

C_{2v}	E	C_2	σ_v	σ_v'
E	E			
C_2				σ_v
σ_v				
σ_v'				

First operation

Second operation

C_{2v}	E	C_2	σ_v	σ_v'
E	E	C_2	σ_v	σ_v'
C_2	C_2	E	σ_v'	σ_v
σ_v	σ_v	σ_v'	E	C_2
σ_v'	σ_v'	σ_v	C_2	E

First operation

The symmetry seen in the entries in this table across the leading diagonal (shown dotted) is a characteristic of Abelian groups.

Example 1; C_2 followed by σ_v

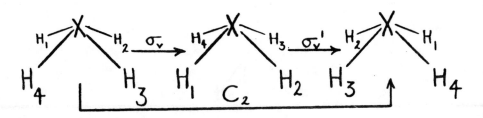

So, C_2 followed by σ_v is equivalent to σ_v'

Example 2; σ_v followed by σ_v'

So, σ_v followed by σ_v' is equivalent to C_2

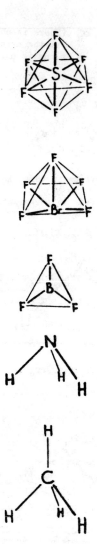

Point Group	Symmetry Operations
C_1	E
C_s	E, σ_h (There is no unique axis of highest symmetry but the axis perpendicular to the mirror plane is unique so σ_h is used)
C_i	E, i
C_2	E, C_2
C_3	E, C_3, C_3^2 (Note C_3^2 means C_3 carried out
C_4	E, C_4, C_2, C_4^3 (twice; C_4^3 means C_4 carried out (thrice etc.
C_5	E, C_5, C_5^2, C_5^3, C_5^4
C_6	E, C_6, C_3, C_2, C_3^2, C_6^5
D_2	E, C_2, C_2', C_2''
D_3	E, $2C_3$, $3C_2$
D_4	E, $2C_4$, C_2, $2C_2'$, $2C_2''$
D_5	E, $2C_5$, $2C_5^2$, $5C_2$
D_6	E, $2C_6$, $2C_3$, C_2, $3C_2'$, $3C_2''$
C_{2v}	E, C_2, σ_v, σ_v'
C_{3v}	E, $2C_3$, $3\sigma_v$
C_{4v}	E, $2C_4$, C_2, $2\sigma_v$, $2\sigma_v'$
C_{5v}	E, $2C_5$, $2C_5^2$, $5\sigma_v$
C_{6v}	E, $2C_6$, $2C_3$, C_2, $3\sigma_v$, $3\sigma_v'$
C_{2h}	E, C_2, i, σ_h
C_{3h}	E, C_3, C_3^2, σ_h, S_3, S_3^5
C_{4h}	E, C_4, C_2, C_4^3, i, S_4^3, σ_h, S_4
C_{5h}	E, C_5, C_5^2, C_5^3, C_5^4, σ_h, S_5, S_5^3, S_5^7, S_5^9
C_{6h}	E, C_6, C_3, C_2, C_3^2, C_6^5, i, S_3^5, S_6^5, σ_h, S_6, S_3

FRAME CONTINUED ON NEXT PAGE

Point Group	Symmetry Operations
D_{2h}	E, C_2, C_2'. C_2'', i, σ_v, σ_v', σ_v'' (The labels on the mirror planes are somewhat arbitrary – one might be labelled σ_h)
D_{3h}	E, $2C_3$, $3C_2$, σ_h, $2S_3$, $3\sigma_d$
D_{4h}	E, $2C_4$, C_2, $2C_2'$, $2C_2''$, i, $2S_4$, σ_h, $2\sigma_d$, $2\sigma_d'$
D_{5h}	E, $2C_5$, $2C_5^2$, $5C_2$, σ_h, $2S_5$, $2S_5^3$, $5\sigma_d$
D_{6h}	E, $2C_6$, $2C_3$, C_2, $2C_2'$, $3C_2''$, i, $2S_3$, $2S_6$, σ_h, $3\sigma_d$, $3\sigma_d'$
D_{2d}	E, $2S_4$, C_2, $2C_2'$, $2\sigma_d$
D_{3d}	E, $2C_3$, $2C_2$, i, $2S_6$, $3\sigma_d$
D_{4d}	E, $2S_8$, $2C_4$, $2S_8^3$, C_2, $4C_2'$, $4\sigma_d$
D_{5d}	E, $2C_5$, $2C_5^2$, $5C_2$, i, $2S_{10}^3$, $2S_{10}$, $5\sigma_d$
D_{6d}	E, $2S_{12}$, $2C_6$, $2S_4$, $2C_3$, $2S_{12}^5$, C_2, $6C_2'$, $6\sigma_d$
S_4	E, S_4, C_2, S_4^3
S_6	E, C_3, C_3^2, i, S_6, S_6^5
T	E, $4C_3$, $4C_3^2$, $3C_2$
T_d	E, $8C_3$, $3C_2$, $6S_4$, $6\sigma_d$
T_h	E, $4C_3$, $4C_3^2$, $3C_2$, i, $4S_6$, $4S_6^5$, $3\sigma_h$
O	E, $8C_3$, $6C_2$, $6C_4$, $2C_2'$
O_h	E, $8C_3$, $6C_2$, $6C_4$, $3C_2'$, i, $8S_6$, $6\sigma_d$, $6S_4$, $3\sigma_h$
I	E, $12C_5$, $12C_5^2$, $20C_3$, $15C_2$
I_h	E, $12C_5$, $12C_5^2$, $20C_3$, $15C_2$, i, $12S_{10}$, $12S_{10}^3$, $20S_6$, $15\sigma_v$

The multiplication table is

C_{2v}	E	C_2	σ_v	σ_v'
E	E	C_2	σ_v	σ_v'
C_2	C_2	E	σ_v'	σ_v
σ_v	σ_v	σ_v'	E	C_2
σ_v'	σ_v'	σ_v	C_2	E

so that substitution gives

	1	1	-1	-1
1	1	1	-1	-1
1	1	1	-1	-1
-1	-1	-1	1	1
-1	-1	-1	1	1

	1	1	1	1
1	1	1	1	1
1	1	1	1	1
1	1	1	1	1
1	1	1	1	1

	1	-1	1	-1
1	1	-1	1	-1
-1	-1	1	-1	1
1	1	-1	1	-1
-1	-1	1	-1	1

	1	-1	-1	1
1	1	-1	-1	1
-1	-1	1	1	-1
-1	-1	1	1	-1
1	1	-1	-1	1

36

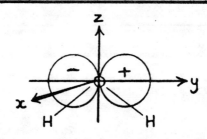

37

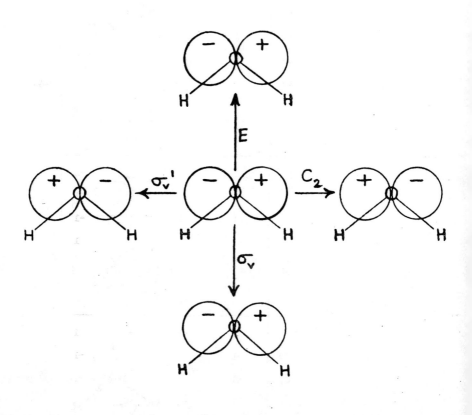

a

b

c

d

e

f

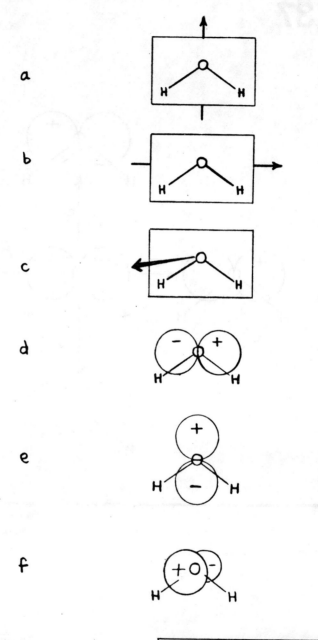

FRAME CONTINUED ON NEXT PAGE

g

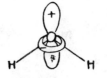

h

i

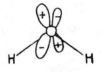

j

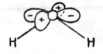

k

l

FRAME CONTINUED ON NEXT PAGE

m

n

o

A _representation_ of a group is a set with the property that the members of the set multiply (using an appropriate law of multiplication – which may be ordinary multiplication, matrix multiplication or some other form of combination) in a way which is isomorphous to the multiplication (i.e. one followed by the other) of the operations of the group.

In the applications with which we are concerned such representations are matrices; in this tape we largely concentrate on 1 x 1 matrices – these are ordinary numbers. Further, it is usually possible to work with the sum of those elements of the matrix which fall along the leading diagonal – the _character_ of the matrix rather than the whole matrix.

When such a set of matrices may simultaneously be reduced to a block-diagonal form we have a _reducible representation_ of the group, when they cannot be so reduced we have an _irreducible representation_. The characters of the matrices of the irreducible representations are listed in the _character table_ of a group.

The _totally symmetric irreducible representation_ of a group has a character of 1 for all operations of the group. It describes the symmetry properties of something which is turned into itself by every one of the operations of the group.

a)

C_{2v}	E	C_2	σ_v	σ_v'
A_1	1	1	1	1
A_2	1	1	-1	-1
B_1	1	-1	1	-1
B_2	1	-1	-1	1

b)

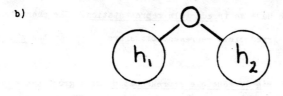

C_{3v}	E	$2C_3$	$3\sigma_v$
A_1	1	1	1
A_2	1	1	-1
E	2	-1	0

C_{2h}	E	C_2	i	σ_h
A_g	1	1	1	1
B_g	1	-1	1	-1
A_u	1	1	-1	-1
B_u	1	-1	-1	1

D_{2h}	E	$C_2(z)$	$C_2(x)$	$C_2(y)$	i	$\sigma(xy)$	$\sigma(yz)$	$\sigma(zx)$
A_g	1	1	1	1	1	1	1	1
B_{1g}	1	1	-1	-1	1	1	-1	-1
B_{2g}	1	-1	1	-1	1	-1	1	-1
B_{3g}	1	-1	-1	1	1	-1	-1	1
A_u	1	1	1	1	-1	-1	-1	-1
B_{1u}	1	1	-1	-1	-1	-1	1	1
B_{2u}	1	-1	1	-1	-1	1	-1	1
B_{3u}	1	-1	-1	1	-1	1	1	-1

FRAME CONTINUED ON NEXT PAGE

D_{4h}	E	$2C_4$	C_2	$2C_2'$	$2C_2''$	i	$2S_4$	σ_h	$2\sigma_d$	$2\sigma_d'$
A_{1g}	1	1	1	1	1	1	1	1	1	1
A_{2g}	1	1	1	-1	-1	1	1	1	-1	-1
B_{1g}	1	-1	1	-1	1	1	-1	1	-1	1
B_{2g}	1	-1	1	1	-1	1	-1	1	1	-1
E_g	2	0	-2	0	0	2	0	-2	0	0
A_{1u}	1	1	1	1	1	-1	-1	-1	-1	-1
A_{2u}	1	1	1	-1	-1	-1	-1	-1	1	1
B_{1u}	1	-1	1	-1	1	-1	1	-1	1	-1
B_{2u}	1	-1	1	1	-1	-1	1	-1	-1	1
E_u	2	0	-2	0	0	-2	0	2	0	0

T_d	E	$8C_3$	$6\sigma_d$	$6S_4$	$3C_2$
A_1	1	1	1	1	1
A_2	1	1	-1	-1	1
E	2	-1	0	0	2
T_1	3	0	-1	1	-1
T_2	3	0	1	-1	-1

FRAME CONTINUED ON NEXT PAGE

O_h	E	$8C_3$	$6C_4$	$3C_2$	$6C_2'$	i	$8S_6$	$6S_4$	$3\sigma_h$	$6\sigma_d$
A_{1g}	1	1	1	1	1	1	1	1	1	1
A_{2g}	1	1	-1	1	-1	1	1	-1	1	-1
E_g	2	-1	0	2	0	2	-1	0	2	0
T_{1g}	3	0	1	-1	-1	3	0	1	-1	-1
T_{2g}	3	0	-1	-1	1	3	0	-1	-1	1
A_{1u}	1	1	1	1	1	-1	-1	-1	-1	-1
A_{2u}	1	1	-1	1	-1	-1	-1	1	-1	1
E_u	2	-1	0	2	0	-2	1	0	-2	0
T_{1u}	3	0	1	-1	-1	-3	0	-1	1	1
T_{2u}	3	0	-1	-1	1	-3	0	1	1	-1

(O_h is the symmetry group of the cube and of the regular octahedron)

Note that for those point groups given above in which i is a symmetry operation the character table blocks into four, the corresponding characters in each block bearing a very simple relationship to each other. This arises from a relationship between the operators listed at the top of the table. Thus, in the O_h table, i is equivalent to E followed by i, S_6 is equivalent to C_3 followed by i etc. (see Frame 14).

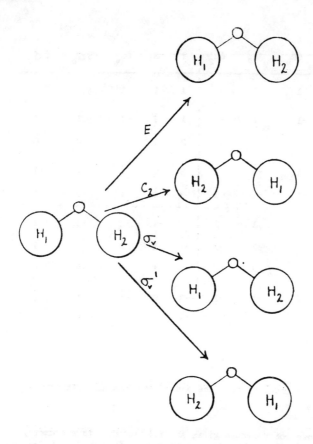

43

a)

C_{2v}	E	C_2	σ_v	σ_v'
A_1	1	1	1	1
A_2	1	1	-1	-1
B_1	1	-1	1	-1
B_2	1	-1	-1	1

and the reducible representation

b)

C_{2v}	E	C_2	σ_v	σ_v'
	2	0	2	0

c)   Select the A_1 irreducible representation; multiply the
characters of the reducible representation by those of the
A_1 irreducible representation and add the products together,

$$(1 \times 2) + (1 \times 0) + (1 \times 2) + (1 \times 0) = 4$$

d)   Divide the result by the order of the group

$$4/4 = 1.$$

The answer, in this case 1, is the number of A_1 irreducible
components in the reducible representation (2,0,2,0).

e)   This is repeated for all the irreducible representations.
Thus for the A_2 irreducible representation

$$(1 \times 2) + (1 \times 0) + (-1 \times 2) + (-1 \times 0) = 0.$$

$$0/4 = 0$$

and we conclude that there are no A_2 irreducible representations
in the reducible representation (2,0,2,0).

FRAME CONTINUED ON NEXT PAGE

f)   For the B_1 irreducible representation

$$(1 \times 2) + (1 \times 0) + (1 + 2) + (-1 \times 0) = 4$$

$$4/4 = 1$$

we have found that there is a B_1 component in the reducible representation $(2,0,2,0)$.

g)   For the B_2 irreducible representation

$$(1 \times 2) + (-1 \times 0) + (-1 \times 2) + (1 \times 0) = 0$$

$$0/4 = 0.$$

That is, there is no B_2 component in the reducible representation $(2,0,2,0)$. Thus, in summary, we have the result that the the irreducible components of the reducible representation $(2,0,2,0)$ are $A_1 + B_1$.

h) Consider the C_{3v} character table

C_{3v}	E	$2C_3$	$3\sigma_v$
A_1	1	1	1
A_2	1	1	-1
E	2	-1	0

and the reducible representation

i)   4   1   0

First multiply these characters by the number of elements in the corresponding classes. Thus

	4×1	1×2	0×3
gives	4	2	0

Now proceed as for the C_{2v} case:-

FRAME CONTINUED ON NEXT PAGE

43
CONTD.

Test for A_1: $(4 \times 1) + (2 \times 1) + (0 \times 1) = 6$

The order of the group is 6, so, $\frac{6}{6} = 1$ (A_1 component).

Test for A_2: $(4 \times 1) + (2 \times 1) + (0 \times -1) = 6$

$\frac{6}{6} = 1$ so there is one A_2 component

Test for E: $(4 \times 2) + (2 \times -1) + (0 \times 0) = 6$

$\frac{6}{6}$ = - so there is one E component.

Thus, the reducible representation $(4, 1, 0)$ has irreducible components $A_1 + A_2 + E$.

C_{2v}	E	C_2	σ_v	σ_v'
A_1	1	1	1	1
A_2	1	1	-1	-1
B_1	1	-1	1	-1
B_2	1	-1	-1	1

C_{3v}	E	$2C_3$	$3\sigma_v$
A_1	1	1	1
A_2	1	1	-1
E	2	-1	0

D_{2d}	E	$2S_4$	C_2	$2C_2'$	$2\sigma_d$
A_1	1	1	1	1	1
A_2	1	1	1	-1	-1
B_1	1	-1	1	1	-1
B_2	1	-1	1	-1	1
E	2	0	-2	0	0

FRAME CONTINUED ON NEXT PAGE

44

1.

C_{2v}	E	C_2	σ_v	σ_v'
	4	2	0	2

2.

C_{2v}	E	C_2	σ_v	σ_v'
	7	-1	1	-3

3.

C_{3v}	E	$2C_3$	$3\sigma_v$
	7	1	-1

4.

C_{3v}	E	$2C_3$	$3\sigma_v$
	3	0	1

5.

D_{2d}	E	$2S_4$	C_2	$2C_2'$	$2\sigma_d$
	4	0	0	-2	0

6.

D_{2d}	E	$2S_4$	C_2	$2C_2'$	$2\sigma_d$
	6	0	2	-2	-2

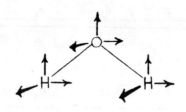

The E operation

Character = 9

The C_2 operation

    the 'after' positions
    of the arrows are
    shown dotted

Character = -1

The σ_v operation

Character = 3

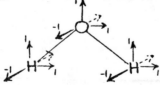

The σ_v' operation

Character = 1

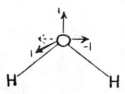

C_{2v}	E	C_2	σ_v	σ_v'
A_1	1	1	1	1
A_2	1	1	-1	-1
B_1	1	-1	1	-1
B_2	1	-1	-1	1

The reducible representation is

	E	C_2	σ_v	σ_v'
Γ_{red}	9	-1	3	1

Test for A_1

	E	C_2	σ_v	σ_v'	
A_1	1	1	1	1	
$\Gamma_{red} \times A_1$	9	-1	3	1	sum = 12; divide by the order of the group (4) $\Rightarrow$ 3. Hence Γ_{red} contains $3A_1$

Test for A_2

	E	C_2	σ_v	σ_v'	
A_2	1	1	-1	-1	
$\Gamma_{red} \times A_2$	9	-1	-3	-1	sum = 4; hence Γ_{red} contains A_2

Test for B_1

	E	C_2	σ_v	σ_v'	
B_1	1	-1	1	-1	
$\Gamma_{red} \times B_1$	9	1	3	-1	sum = 12; hence Γ_{red} contains $3B_1$

Test for B_2

	E	C_2	σ_v	σ_v'	
B_2	1	-1	-1	1	
$\Gamma_{red} \times B_2$	9	1	-3	1	sum = 8; hence Γ_{red} contains $2B_2$

FRAME CONTINUED ON NEXT PAGE

47
CONTD.

Hence Γ_{red} contains $3A_1 + A_2 + 3B_1 + 2B_2$

From the solution to problems a, b, c

and m, n and o of Frame 38 (given

in Frame 57) we conclude that:-

a) The translations of H_2O transform as $A_1 + B_1 + B_2$

b) The rotation of H_2O transform as $A_2 + B_1 + B_2$

Subtract these from the components

of Γ_{red} to obtain the symmetry species

of the vibrations of H_2O. These are $2A_1 + B_1$

48

C_{2v}	E	C_2	σ_v	σ_v'		
A_1	1	1	1	1	z , T_z	x^2 , y^2 , z^2
A_2	1	1	-1	-1	R_z	xy
B_1	1	-1	1	-1	y , T_y , R_x	yz
B_2	1	-1	-1	1	x , T_x , R_y	xz

The B$_1$ mode (schematic)

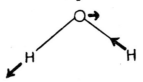

The A$_1$ modes (schematic)

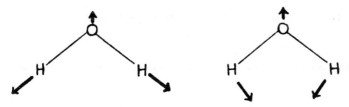

Because the molecule must not rotate (rotations have been
factored out) the H atom motions must lie in the yz plane for both
of the A$_1$ modes. Further, their motions in the two A$_1$ modes must
be quite different (the two A$_1$ modes must be quite different - they
must be orthogonal). It makes chemical sense that one mode should be,
essentially, an O-H stretching mode. It follows that the second A$_1$
mode must have the general form shown.

The A_2 irreducible representation of the C_{2v} point group is

	E	C_2	σ_v	σ_v'
A_2	1	1	-1	-1

The B_1 irreducible representation is

B_1	1	-1	1	-1

Form the direct product by multiplying pairs of characters together

$A_2 \times B_1$	(1x1)	(1x-1)	(-1x1)	(-1x-1)	
	1	-1	-1	1	$= B_2$

We conclude that the direct product $A_2 \times B_1$ is B_2.

The <u>direct product</u> of two representations is the representation
obtained when pairs of corresponding characters are multiplied
together

The direct product table of the C_{2v} point group

C_{2v}	A_1	A_2	B_1	B_2
A_1	A_1	A_2	B_1	B_2
A_2	A_2	A_1	B_2	B_1
B_1	B_1	B_2	A_1	A_2
B_2	B_2	B_1	A_2	A_1

The direct product table of the C_{3v} point group

C_{3v}	A_1	A_2	E
A_1	A_1	A_2	E
A_2	A_2	A_1	E
E	E	E	A_1+A_2+E

Note that the totally symmetric irreducible representation (A_1 in both of the
above tables) only appears on the leading diagonal of these tables. This
is invariably the case for all direct product tables. We conclude that
it is always true that:-

> The totally symmetric irreducible representation is
> only generated when an irreducible representation is
> multiplied by itself.

52

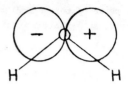

53

Oxygen orbital	Symmetry species	Does integration give a non-zero result?
s	A_1	Yes
p_x	B_2	No
p_y	B_1	No
p_z	A_1	No
d_{z^2}	A_1	No
$d_{x^2-y^2}$	A_1	No
d_{xy}	A_2	No
d_{xz}	B_2	No
d_{yz}	B_1	No

$$\int \psi_e(^{A_1}) \; \hat{\mu}_x \; \psi_g(^{A_1}) \; d\tau$$

$\hat{\mu}_x$ transforms as B_2 so we have to form the triple direct product

$A_1 \times B_2 \times A_1 = A_1 \times (B_2 \times A_1) = A_1 \times B_2 = B_2$

Integration over all space of a non-totally symmetric irreducible

representation gives zero so we conclude that the A_1 vibration is

not active in x polarization.

$$\int \psi_e(^{A_1}) \; \hat{\mu}_y \; \psi_g(^{A_1}) \; d\tau$$

We have $A_1 \times B_1 \times A_1 = B_1 \Rightarrow$ zero integral. The A_1 vibration is not allowed in

y polarization.

$$\int \psi_e(^{A_1}) \; \hat{\mu}_z \; \psi_g(^{A_1}) \; d\tau$$

We have $A_1 \times A_1 \times A_1 = A_1 \Rightarrow$ non-zero integral. The A_1 vibration is allowed (and

may hence be identified) in z polarization.

$$\int \psi_e(^{B_1}) \; \hat{\mu}_x \; \psi_g(^{A_1}) \; d\tau$$

We have $B_1 \times B_2 \times A_1 = A_2 \Rightarrow$ zero integral. The B_1 vibration is not allowed in

x polarization.

$$\int \psi_e(^{B_1}) \; \hat{\mu}_y \; \psi_g(^{A_1}) \; d\tau$$

We have $B_1 \times B_1 \times A_1 = A_1 \Rightarrow$ non-zero integral. The B_1 vibration is allowed (and

may hence be identified) in y polarization.

$$\int \psi_e(^{B_1}) \; \hat{\mu}_z \; \psi_g(^{A_1}) \; d\tau$$

We have $B_1 \times A_1 \times A_1 = B_1 \Rightarrow$ zero integral. The B_1 vibration is not allowed in

z polarization.

The A_1 vibrations are

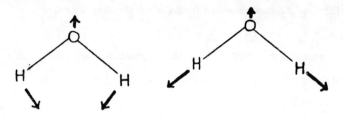

In both cases the dipole moment changes along the z direction
and it is only the integral

$$\int \psi_e^{(A_1)} \hat{\mu}_z \, \psi_g^{(A_1)} d\tau$$ which is non zero (direct product $A_1 \times A_1 \times A_1 = A_1$)

The B_1 vibration

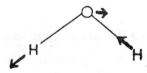

The dipole moment changes along the y
direction and it is only the integral

$$\int \psi_e^{(B_1)} \hat{\mu}_y \, \psi_g^{(A_1)} d\tau$$ which is non-zero (direct product $B_1 \times B_1 \times A_1 = A_1$)

SF_6	O_h
BrF_5	C_{4v}
BF_3	D_{3h}
NH_3	C_{3v}
CH_4	T_d

57

	E	C_2	σ_v	σ_v'
a) Translation along z	1	1	1	1
b) Translation along y	1	-1	1	-1
c) Translation along x	1	-1	-1	1
d) The oxygen p_z orbital	1	1	1	1
e) The oxygen p_y	1	-1	1	-1
f) The oxygen p_x	1	-1	-1	1
g) The oxygen d_{z^2}	1	1	1	1
h) The oxygen d_{yz}	1	-1	1	-1
i) The oxygen d_{zx}	1	-1	-1	1
j) The oxygen $d_{x^2-y^2}$	1	1	1	1
k) The oxygen d_{xy}	1	1	-1	-1
l) The dipole moment	1	1	1	1
m) Rotation about z	1	1	-1	-1
n) Rotation about y	1	-1	-1	1
o) Rotation about x	1	-1	1	-1

1. $2A_1 + A_2 + B_2$

2. $A_1 + 2A_2 + 3B_1 + B_2$

3. $A_1 + 2A_2 + 2E$

4. $A_1 + E$

5. $A_2 + B_2 + E$

6. $2A_2 + B_1 + B_2 + E$